SPIDERWEBS & SPIDER SILK

Let's Take a Closer Look

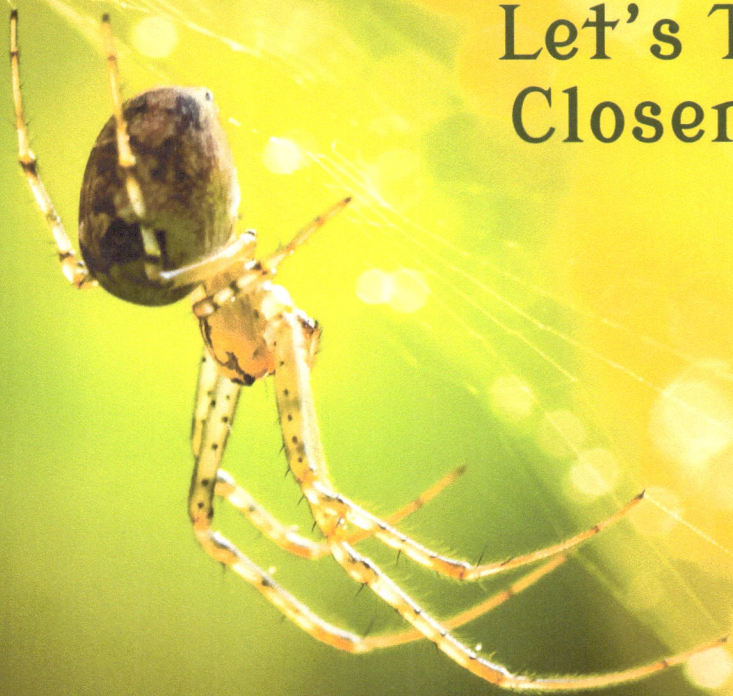

Lynnae W. Allred

To my grandchildren,
who ask insightful questions.

Special thanks to Dr. Randy Lewis and Don Drife for their
expertise and encouragement. They made this a better book.

ISBN 13: 978-1-7379746-0-4
ebook ISBN 13: 978-1-7379746-1-1

Printed in the United States of America

10 9 8 7 6 5 4 3 2 1

Printed on acid-free paper

For more information, email customercare@playdatebox.com

A spiderweb is an amazing creation! And spider silk is so versatile that spiders have discovered thousands of interesting ways to use it.

HOW DO SPIDERS MAKE SILK?

Spiders produce a liquid gel inside of their bodies made of protein. Your hair is made out of protein. If you think about it, your hairs look a little like the strands of a spiderweb. Spider silk is even finer than human hair.

The protein gel inside the
spider's body flows into a
narrow tube where it is
acidified and stretched.
This forms a fiber.
Organs at the back of
a spider's body, called
spinnerets, control how
fast the silk comes out
of the spider's body, and
how thick or thin it is.

Spiders can make up
to seven different
kinds of silk: thick
or thin, sticky or
scented, colorful and
reflective, or nearly
transparent.

CAPTURING FOOD

Many spiders use their silk
to build webs to help them
capture food. Orb-weaving
spiders capture their food in
sticky spiderweb nets. Some
build a new web every day.

This is a funnel web. The spider who built this web is hiding in the opening at the bottom waiting to ambush its prey.

A bowl and doily web is another unusual web shape. Suspended above a bowl-shaped web are many nearly invisible strands of spider silk. A flying insect accidentally crashes into one of these strands and falls into the bowl. It's dinnertime for the spider!

Net-casting spiders hold onto a piece of web with their four front legs and throw it over their prey like a net.

A slingshot spider will create a twisted spring out of spider silk that allows it to catapult itself and its web toward its prey at ten times the speed of a fighter jet.

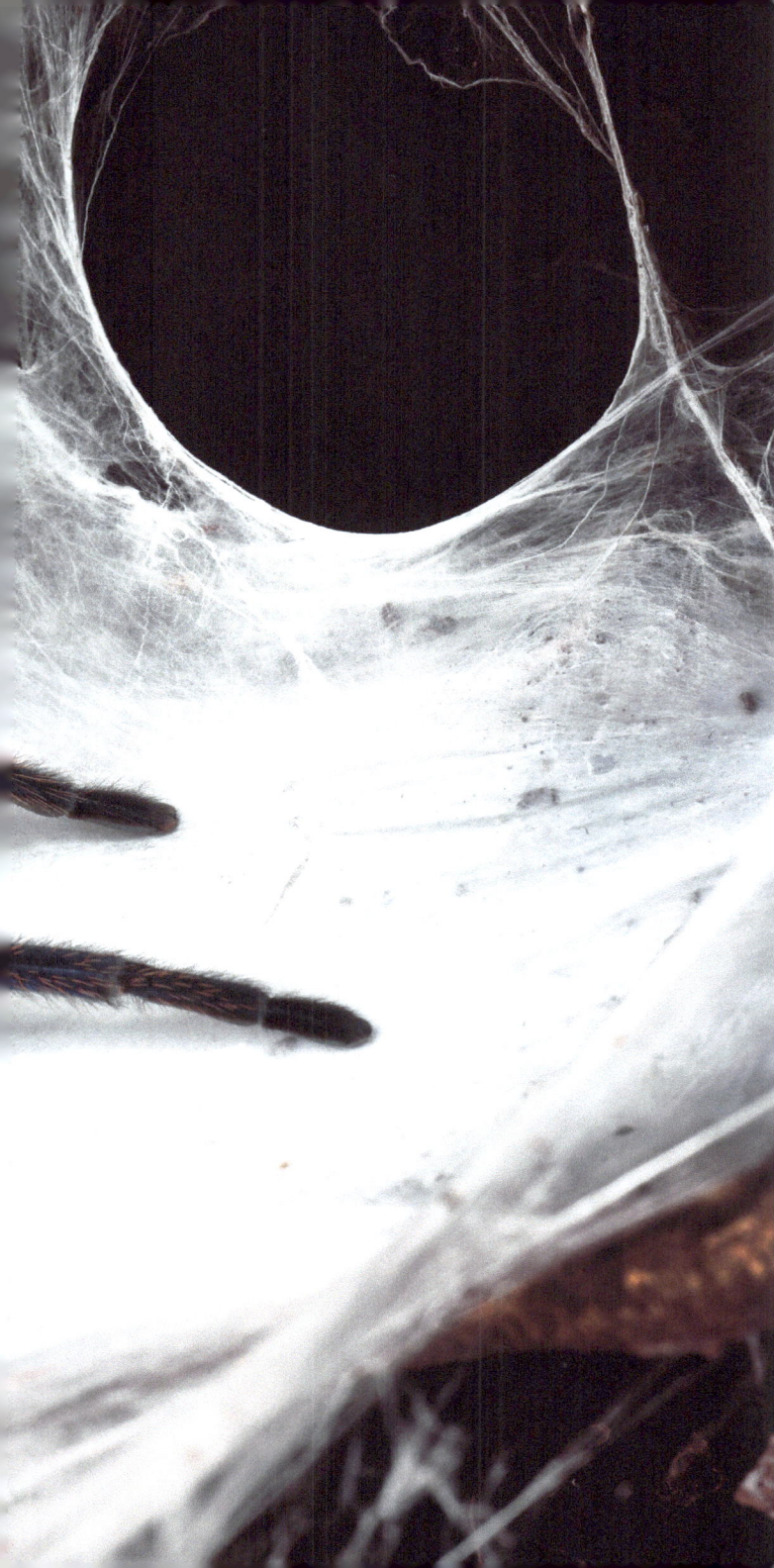

SHELTER

Some spiders don't build webs for catching food. They prefer to hunt for their food and use their spider silk in other ways. A tarantula uses spider silk to line and protect its burrow.

Other spiders make shelters of sticks and leaves knit together with spider silk.

A diving bell spider builds a special dome of silk and fills the dome with air bubbles so that it can live underwater.

GETTING AROUND

Spiders can also use silk to transport themselves from one place to another. Sometimes baby spiderlings and even adult spiders will lift their abdomens into the breeze and spin many strands of silk all at once. The wind lifts the spider into the air so it can travel like a hang glider. This is called ballooning.

A spider might use a single dragline of silk to help it find its way back home again. It can also spin a dragline for dropping down quickly to capture prey. Or, it may drop out of sight on a dragline to escape from a predator.

WRAPPING THINGS

Spider silk can also be used to wrap up an insect after it has been captured. The spider will eat the insect later. Male spiders sometimes wrap up dead bugs to give to female spiders as gifts.

A mother spider can use silk to wrap up her spider eggs into a pouch or ball to keep them safe. These eggs will hatch into more spiderling babies. Some spiders spin a pouch with a soft inner lining to cushion the eggs. Then they cover that with a second stiff, hard silk to protect the eggs.

CAN HUMANS MAKE SPIDER SILK?

Humans have worked for many years to create spider silk in a lab because by weight, spider silk is stronger than steel and more elastic than nylon. It is incredibly flexible, thin, and light.

Imagine if you could twist many strands of spider silk together into a rope as thick as your finger. Next, you could weave the yarn into a net. Your new net would be strong enough to halt a speeding subway train!

Scientists hope that someday when people can make spider silk as efficiently as spiders, we will invent thousands of ways to use it: A shirt as light as a pencil that could stop a bullet, a bandage to repair a wound, or socks that would never wear out!

RECYCLING OLD SPIDER SILK

When a spider is finished with her web, she will often eat it. Why would a spider do that?

When a spider eats an old web or an insect, it can digest
the insect or used web and turn it into liquid protein.
And if you have liquid protein, you can make
something really amazing out of it.

MORE
SPIDER SILK!

Scan the code on the right with the photos app on your smartphone to see videos of spiders building webs, wrapping up live bugs, and carrying a whole nursery of eggs around with them.

Visit playdatebox.com/spiders for a spider pizza recipe and other fun spiderweb activities.

www.ingramcontent.com/pod-product-compliance
Lightning Source LLC
Chambersburg PA
CBHW061153030426
42336CB00002B/31